SUSTAINABLE SUPPORT

SUSTAINABLE SUPPORT

SUSTAINABLE SUPPORT

BLOW YOUR CUSTOMER'S MIND
(WITHOUT LOSING YOURS)

BY

BEN MEREDITH

Library of Congress Control Number: 2024927546

Epub ISBN 979-8-9922995-1-9
Paperback ISBN 979-8-9922995-0-2

INTRODUCTION

Customer Service is hard. Technical Support is hard. Annual turnover for technical support positions is as high as 40%, according to leading Technical Support expert Jeff Rumburg. That means that four out of ten of your current technical support team will not be here this time next year. Agent turnover is incredibly costly for companies, exhausting for team morale, and—believe it or not—mostly preventable.

Management should be concerned about this churn, and motivated to find its root cause.

Worse, in watching that churn we've resolved to watch our customers take to social media and lambast us for having technical support that is neither technical nor supportive—and they're not wrong.

You're here because you know something's broken and there's got to be a way to do Customer Service and Technical Support in a way that makes it more than a loss on the P&L sheet.

In this book, I'm going to get practical on how to fix Technical Support, but not before I get philosophical. Our problem is way more than practical. From CEO to freelancer to manager to marketer, we all have been sold a busted theory about technical support, and instead of examining it, we're just resolved to watch perfectly capable humans run out to find their next job after suffering through a few months of working in Customer Support.

Technical support can be fun, engaging, and rewarding work.

Customers can come away from an experience with Technical Support that leaves them encouraged and genuinely helped.

If you're a support agent, you'll get some practical help here, but the best reader of this book is a support supervisor, manager, or director who wants a proven path forward to set your team up for success, avoid burnout, and create a system for blowing customers' minds without losing yours.

INTERLUDE: MY STORY

I graduated from the University of North Carolina at Chapel Hill with a degree in religious studies. My focus was early American religious history. I lovingly referred to my degree as "pre-unemployment" and quickly learned that others in my department—most specifically the professors—did not share my sense of humor.

Lest this introduction prompt you to flip back to the cover to assure yourself what book you are reading, I tell you all of that because (perhaps not curiously) none of my classes or formal training involved CSS, HTML, PHP, or coding your own website.

If I were to tell you the *year* I graduated it would become even more evident that, on paper, it's something I have essentially no credentials for. Suffice it to say that the entire field of self-publishing on the web during my undergrad years did not overlap with people outside of the fledgling computer science departments. Computers were barely personal and phones were not smart (not to mention mostly attached to walls).

My utter lack of qualification makes what happened in 2008 even less likely. Sitting around after a staff meeting at the faith-based nonprofit I worked for, my boss casually mentioned "You just got a new computer, maybe you could create us a website!"

As a fun aside: that statement has now come full circle, because now when I mention that I lead teams of web software technical support agents it quickly draws comments from even my newest acquaintances of "oooh, can you fix my printer?" I... rarely know how to respond to that question.

To be fair to that boss, I did want to get more into digital creation, and so I readily agreed and set about making a website. I had heard of WordPress, and a friend suggested a premium theme I should check out. I think his exact words were "This makes it easy to create a site."

Spoiler alert: it was not easy.

Included in the purchase of that premium theme was access to a support forum where I was able to ask questions when and if I got stuck trying to make something happen.

I did indeed get stuck, early and often. That was the very moment I made a subconscious and unintentional career shift (without even leaving my job!): I started writing on and participating in the technical support forums, mostly as a bumbling user, and then as one trying to help other bumblers.

A few years later, in a case of the universe making me cash in on my teenage humor, I began using my "pre-unemployment" degree, as a member of the job-seeking class.

While looking for a full-time gig I started freelancing in web maintenance on the side. I was by that time a sort of digital generalist, dabbling with code and tinkering with websites.

I got a primary client who was helping the ends to meet, and a handful of other clients at the same time. I created an LLC and started to treat my little hustle with full-time energy. Some of those clients remain with me even today. I devoured enough freely available training in PHP to launch a WordPress plugin, with plans to use it to prove (a little bit to myself but mostly to the watching world) that I could support a product.

That plugin—Better Click to Tweet—ended up being the résumé to get me a job in product technical support. For 5 years I worked in the trenches of tech support, slogging through support tickets, and communicating with the team.

Something strange happened: in an industry that has a frankly abysmal churn rate for team members, and an earned reputation for making agents generally miserable, I really enjoyed doing technical support.

This immediately raises three questions:

1. Is it really possible to *enjoy* technical support?

2. Do you have to be some specific personality type to enjoy technical support?

3. Can technical support be more than a cost for your organization?

The principles outlined and explained in this book will answer these questions more thoroughly, but in short: yes, no, and yes.

Over the first few years of daily slogging through email- and forum-based technical support, my mentor and boss Matt Cromwell and I started to notice a handful of foundational principles, and write them down.

More than that, I've found these principles to be transferable: during those 5 years, I was promoted first to be a supervisor, then a manager, and now a Director of technical support. As Matt and I set about training new team members, we wanted to see if the fledgling principles that were so effective for us personally could scale to entire teams.

Never short on optimism, we set out.

Five separate teams of technicians (all happily performing better than they ever have) later, and we have our answer: You can do technical support in a sustainable way.

It's possible to blow your customer's mind, and not lose yours.

Oh yeah, and another full-circle moment: I've discovered that technical support is only sustainable by informing and infusing it with principles I learned and wrote about in stuffy religious studies classrooms. Beneath the technical problem there's a person, and our primary job in technical support is to support and empower the person.

So maybe I was trained for this all along!

SUSTAINABLE SUPPORT

Part One

Old School Support

CHAPTER ONE: STEVE'S TROUBLE

Scenario: you are a technical support agent for a distributed web-based software product and a customer (let's call him Steve) reaches out frantically via email support:

> *Hey, my website is down and all visitors are greeted with "There has been a fatal error on this site" and I don't know what to do. I just sent an email to 2500 people announcing a sale, and they are starting to email me to let me know that they can't get the deal. Please help. This is costing me hundreds of dollars every hour!!*

How should you respond?

What does Sustainable Support of Steve look like? How can we ensure that our response moves him toward resolution from the very first word? What things do we need to avoid and leave out of the responses?

The problem with the usual approach to technical support specifically as well as customer service more generally is that we default to a posture that errs in one of two directions: we're either unintentionally confrontational or fake-sympathetic. Before we get into an overarching way to label (and therefore spot and eliminate) the bad approach, let's pause to examine those two errors a bit.

Accidentally using Fighting Words

Unintentionally confrontational support is an honest mistake: you're trying to do your job well, and to instruct users how to best help you do your job. In Steve's case, he's reached out to the wrong team! This is a server-level problem, most likely, so we need to skip straight to sending him to his web host, right? Confrontational support ensures that we only help him once we've determined that the problem is indeed with our product.

Using Empty Platitudes

Fake-sympathetic support looks like "I'm so sorry you're going through this... here's 26 steps that may or may not fix it." It's possible that you are actually sympathetic and you do actually feel bad for the user, but saying (as opposed to demonstrating) it is a sure-fire way to come off as fake. More on that later.

Both of these extremes are examples of what I call "Old School Support." The best way to really understand the problem is an example: let's take a look at an Old School answer to Steve:

> *So sorry you're experiencing this!*
>
> *That's a fatal error on the site, and I can't really help with that.*
>
> *It usually means that there's a server-level conflict between two plugins, or a plugin and the theme, or between some code on the site and the underlying PHP version.*

You'll need to contact your web host to get them to let you back into the site. Let me know what they say, and I'm happy to give you some more direction.

What is wrong with that answer?

You might be reading that and thinking "hey, that's not a bad answer!" After all, it's not rude or condescending, it's direct and explains fairly well that they have reached out to the wrong support team, and (bonus!) it explains the technical concepts!

What makes this response "Old School" is that it moves Steve away from resolving his problem, by being unintentionally confrontational and faux-sympathetic. There are 5 specific ways that it fails to move Steve toward resolution.

1. Leading with apologies
2. Technical jargon
3. Leading with a negative
4. Passing the buck
5. Giving them homework

Let's unpack each one:

1. Leading with Apologies

In the first few words, our Old School reply set the exact wrong tone: one of apology and lack of confidence. The thing to keep in mind with apologies is that they do not communicate confidence. Right out of the gate, the user is thinking "oh no, I've come to the wrong place!" instead of what we want: for them to be empowered to solve their problem.

2. Technical Jargon

By using technical jargon like "fatal error" and "server-level conflict" you're potentially derailing them with concepts they don't yet understand, and moving them toward making the problem even harder to solve.

3. Leading with a Negative

Similar to apologizing, starting out with "I can't help with that" is setting the customer up to be frustrated. Even (and especially) when it's true that you can't resolve their problem with one quick reply, you can *always* help. It's an Old School impulse to think you can't.

4. Passing the Buck

Two "help desks" trapping customers between them and pointing fingers at each other is sadly common. I'll argue later in the book that this is not even helpful for you as a technical support agent, because in an effort to "only support our product" you're creating a (repeated) opportunity to view yourself as less than an "agent," but for this point, focus on how unhelpful buck-passing is for Steve, who is stuck just wanting *someone* to help.

5. Giving them homework

By telling Steve to "contact {the} web host" you've sent him away to do work, but not equipped him to really understand what he's doing. Like a teacher giving him a pop-quiz on a topic they've never covered, this sends Steve into a world where at best he's just parroting words from you to the other support team. You've positioned him as the expert, but failed to give him any of the expertise.

These five mistakes are not the only ones you can make in technical support, as of course you can do egregious things like giving incorrect answers, or being rude, or somehow otherwise being incorrect in philosophy and approach.

These five errors are exceedingly common among support agents and supervisors who—get this—*believe themselves to be doing a good job*!

We've all fallen victim to the false belief that it's just supposed to be a generally terrible thing to have to talk to and support customers, so we've assumed that the normal friction and frustration is just part of the job.

It's not.

With practice you'll start to see each one of these five errors in your own tickets. But before we have any hope of a sustainable system of customer support, we'll need to identify the philosophy undergirding Old School support.

CHAPTER TWO:
SILENT PHILOSOPHY

Carbon Monoxide (known by its chemical abbreviation CO) is a poisonous, odorless gas produced as a byproduct of burning coal or gas in engines, appliances, and other processes.

Since it's odorless you can unknowingly poison yourself. This perfectly natural element in our atmosphere is essentially plotting to kill you, via the fancy gas appliance in your kitchen.

If some impossible-to-detect force is plotting against you, the only logical solution is to build a system to detect it. In the case of Carbon Monoxide, that's a small detector that costs less than a nice meal, plugged in and ready to alert you when CO levels get out of hand.

For detecting Old School Support, consider this chapter your built-in alarm system.

Old School Support is based on at least three faulty premises. With practice, you'll be able to spot these tendencies in your own mindset and the customer interactions of agents on your team.

In many ways, the premises are only detectable by their results. The premise itself operates like a silent and odorless gas. We're giving the odorless gas of Old School Support a distinctive smell for you to be able to pinpoint and root out the causes.

It might not feel practical to talk philosophy, but until you clear out these mental errors (reinforced by decades of experience on both sides of technical support) you'd be better off trying to drive a car down a road submerged under water.

Before we dive into what makes Sustainable Support stand out, we first need to recognize and challenge three common faulty premises.

Faulty Premise 1: You can make a product so good that it doesn't need support.

There's no clearer example of this premise in action than the announcement of the first iPhone.

Those of us old enough to remember phones in 2006 can clearly recall opening the box, taking out the tiny coffin-shaped brick, and then grabbing the equally-thick user manual, written in at least 3 languages, and spanning 25-30 pages per language.

The manual taught you in step-by-step fashion how to use your new device: how to set up the home screen, how to set the clock, how to access the tiny-but-addicting game of snake that came preloaded. It was a useful guide to your new device.

So in 2007, standing professionally-lit on a stage in San Francisco, Steve Jobs made a bold and startling claim. He held the new iPhone above his head and declared with a grin "It doesn't even need a manual!"

And in one stroke, he made both a brilliant marketing move, and underscored a blatant lie. This new product is so good, so intuitive, so simple that it doesn't need anyone to teach new users how to use it.

Should you care to prove that it's a lie, I'd invite you to place a bookmark here, and then visit your nearest Apple Store. Stand close enough to the "Genius Bar" at the back of the store, and discreetly count how many folks are there for help with their iDevices.

See, it's just *not true* that you can create a device, piece of software, or experience so good that a first-timer won't need help with it.

But wow do we as an industry desperately want the Prophet Steve to have been correct. Decades later, it's time to admit it: humans are collaborative by nature, and there's no such thing as a product, service, software, or experience that will dampen our impulse to ask for help understanding and using it.

The realization that the premise is faulty allows you to start reframing what support is, due in large part to the structures and worldviews we've built on top of that faulty premise:

Result 1: Every ticket an indictment

If it's possible to make a product so good that new users won't need support, then a pile of 25 (or 100, or 1,000) tickets when you open your helpdesk software on Monday morning becomes a silent problem. Each ticket is reinforcing to you that your product is not good enough. Like an injury attorney in an expensive suit, your inbox is building a case against you and your product.

Result 2: Inbox zero tunnel vision

There's a real temptation (if it's true that a utopia exists where products are so good they don't need support) to get laser-focused on inbox zero in support. If all those tickets are little (and big!) prosecuting attorneys, then what we want is some loophole to get the case dismissed.

To hang on a bit longer to the legal metaphor: this faulty premise creates a desire and motivation to hide the evidence. Managers and directors can focus far too much on hitting inbox zero.

Result 3: Agent anxiety

If you subconsciously believe that inbox zero is the goal, and that every ticket is charging your product team with "guilt," it is a fairly short journey to burnt-out and disengaged agents.

Until you start preaching to yourself and your team that tickets (even lots of them) are a good thing, you run the risk of driving your best and most engaged agents away.

Premise 2: Tech support is for junior-junior developers.

Ask even the least informed member of the public the difference between a software engineer and a support technician, and they will nearly universally make the same mistake: they'll compare skill levels of the two titles.

It's an honest mistake, as a brief perusal of job descriptions and salary ranges will quickly show that engineers and support agents reside on different ends of the value spectrum. So naturally you could infer that the discrepancy of value means that the skill levels of engineers are the primary differentiator.

Here's the reality: comparing the skill level of engineers and support agents is like comparing the skill level of a baseball player with the skill level of the folks who tend the baseball field, planting and watering grass, sanding and raking the pitcher's mound, etc.

The more experienced and knowledgeable about the game of baseball the groundskeepers are, the better they can facilitate and create an environment for the baseball players to thrive. But it's wise to notice that at a certain point being a high level groundskeeper is a fundamentally different set of skills than swinging a bat or timing a stolen base.

Technical support agents are consistently viewed as essentially "engineers, but not good ones." This is often reinforced by labeling folks "tier 1," "tier 2," etc.

Since there's an overlap in knowledge (and the best support agents have a high level of knowledge about technical concepts) it's assumed that there will also be an overlap in skills.

The skills of a support agent are diagnostic, triage, and assimilation of data into replicable steps for the engineers to fix. The skills of an engineer are to debug, patch, and resolve issues—not to mention develop and create new solutions!

The two roles are highly complementary.

When you mistake a support agent to be essentially a junior-junior engineer, that leads to 3 predictable outcomes.

Result 1: Passing the buck to other support desks

If support agents are (viewed as) developers who haven't grown up into real valuable ones yet, it makes it a foregone conclusion that when they run into problems that are not directly related to their skills, there's a real tendency to pass that problem along to some other support desk.

Passing a user along to another support desk can be done correctly (more on that later). But the impulse to say "not my problem" is largely expected when the organization and managers treat support agents like lesser-developers.

Result 2: Disengaged technicians

That leads into the next result, directly: it's not only a risk that agents will disengage from issues that are outside of scope; there's a significant risk that they will disengage from issues that are directly related to your product.

By putting a support technician in a drawer labeled "technical, but not very" you've capped off their potential, and invited them to look for other jobs where they can perform (and be compensated) at a higher level.

Result 3: Internal isolation

Not only are they disengaged by that mindset, they are also less engageable. Other teams are not going to reach out to the technical support team for answers to questions, if they'd be better suited to just go directly to the experts (the engineers!).

It's fairly dehumanizing to be treated like "just the mouthpiece for the smart people." After all, that's a job a trained chatbot can do, so it's *predictably* dehumanizing.

Premise 3: Technical Support is done-able.

Back to quizzing the general public, but this time about Firefighting. Ask your standard muggle what a firefighter does, and you'll likely get various answers that all centralize around what we see firefighters doing: they go to where fire is, and they put it out.

I'm sure there are things other than "pointing the hose at the fire" that good firefighters do. But I don't know about those things.

Too many of us see technical support agents in the same one-dimensional way: like firefighters who do nothing more than point a hose at the fire.

The ticket queue is the fire, and once there are no tickets (see desired result of faulty premise 1) we assume that they are done. In fact, that's the faulty premise at work: we think technical support is a finite and one-dimensional thing that you can "finish."

Result 1: No concern for churn

If tickets are fires, and tickets come from customers, and an easy way to reduce fires is to make customers go away, a "hose pointer" is more than happy to see folks leave. You've nearly incentivized it at that point.

Fewer customers means fewer fires.

Result 2: Obsessing over Time to First reply

If support is nothing more than a cost, and something that can be checked off as "finished" then management can easily fall into a pattern of measuring effectiveness using the wrong metric.

The premier among wrong metrics is the Time To First Reply (TTFR). Of course we want to reply as effectively as possible to tickets, chats, etc. But Sustainable Support cares more for "effective" than "fast" because it's much too narrow a picture to care more about how fast they heard from us, when we could focus on how fast their issue was fully resolved.

Result 3: Heavy lines around scope

The final result of the "hose-pointer-only" faulty premise is that your agents are more likely to punt to other helpdesks. Establishing the agent's job as "put out fires" makes them more likely to carefully map out "fire districts" where they can say "That's someone else's fire!"

Rooting out the Carbon Monoxide

Before you move on, do a simple sweep to root out Old School support philosophy, by flipping each premise on its head. Take stock of each of the nine results listed in this chapter, treating it like the beeping of an Old School support detector plugged into the wall of your kitchen. Here's the checklist:

- ✓ Tickets feel like indictments
- ✓ A focus on Inbox zero
- ✓ Anxious agents
- ✓ Passing the buck
- ✓ Disengaged agents
- ✓ Internal agent (and support team) Isolation
- ✓ Heavy lines around Scope
- ✓ Obsessing over TTFR
- ✓ Churn that doesn't bother agents

Each of these outcomes is a sure sign that you have some Old School philosophy lurking in your organization. Left alone, it's silently poisoning your Sustainable Support system.

Chapter Three:
Setting the Teams

Here in the southern United States, the word "y'all" is a perennial favorite, universally used across an entire swath of the country. Among the reasons: it's inclusive, easy to say, and part of our heritage.

To an outsider, "y'all" might come across as just a lazy way of saying "you all" or a sign that the speaker is less articulate. Even if that is true, it misses the point. The real magic of "y'all" is one layer deeper than that: the concept of "all y'all."

"All y'all" is a turn of phrase best coupled with a specific movement of your arms, where you sweep out in large circles at the neck level, like you're trying to give the whole room a hug. We loop in "all y'all" for special things like being invited to the cookout or going to the beach. *All y'all are welcome.*

"All y'all" is also more than a turn of phrase. It's a mindset toward your fellow human that breaks down walls, makes adversaries teammates, and finds common ground. *All y'all* are welcome.

The three faulty premises from chapter 2 conspire to silently poison your support team, and underneath it all, that death looks like the functional opposite of "all y'all": Old School Support separates all the groups into teams all arguing with each other about whose problem it is.

Old School support has the end user carrying the problem around to various help desks all arrayed against one another (and more importantly, aimed against the user themselves). Of all the actors in the story, the user is the one least equipped to understand the problem, and yet the Old School support model has incentivized them having to be frustrated enough to escalate it past the support technicians to someone who cares.

The Nasty Competition of Old School Support

Old School support creates multiple small teams all arrayed against one another, with the user (and their problem) stuck in the middle.

Your User

Sustainable support, on the other hand, creates two teams:

1. All y'all (the user, the other support desks, the web host, the third-party developers, and anybody standing within earshot)

2. The problem.

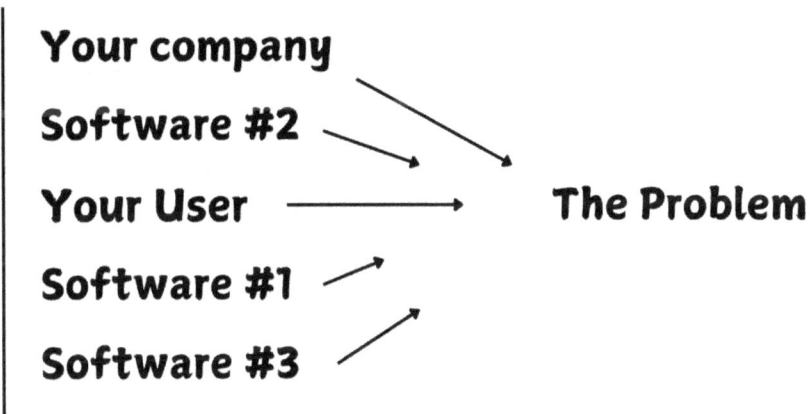

All Y'all

Your company

Software #2

Your User   The Problem

Software #1

Software #3

The point that we are driving toward is simple: your customers are happier when your team is happier, and your team is happier when they are empowered to truly solve problems in a way that arrays everyone involved toward the problem. Until you get the teams right, you're fighting an Old School (losing) battle.

The first step to getting on the same team? Understanding what Sustainable Support is.

SUSTAINABLE SUPPORT

Part Two

What is Sustainable Support?

INTERLUDE:
WHERE WE'RE GOING

The Old School approach to technical support is inherently unsustainable. It is based on faulty premises that lead to "teams" that are all fighting each other instead of the problem.

Before we get into how to build a Sustainable Support system, it is critical to understand what exactly we are talking about. It's tempting to skip ahead to tactics or strategies, but they will only work if we build them on a foundation of understanding. In fact, the deeper your understanding is, the more efficiently you can tactically operate.

The actions of Sustainable Support flow best from a deep grasp of concepts.

So what is Sustainable Support?

In short: Sustainable Support is a *symbiotic* system that *holds in tension* the needs of the *user*, *company*, and *support agent* to *proactively and reactively* resolve technical problems.

A Symbiotic System

The Sustainable support system creates a relationship between all of the departments in the company. In Old School support, decisions flow only in one direction: some other department makes decisions and the support team communicates them. In Sustainable support, insights from the customer are vetted through the support team, validated and communicated by agents who have true *agency*.

That Holds in Tension User, Company, and Agent needs

An agent who is incentivized by Old School thought processes focuses on one area to the exclusion of others. Sustainable support keeps all three areas in proper tension. We want to minimize refunds (user needs), maximize profit (company needs), AND create an environment that reduces burnout and churn (agent needs).

To Proactively and Reactively resolve technical problems

A Sustainable Support system is one that solves every problem that comes across the helpdesk, but also one that seeks to take user feedback and prevent the problems coming across in the first place.

In part two, we'll dig more deeply into the philosophical underpinnings of Sustainable support.

Chapter Four:

Functional Premises of Sustainable Support

Just as Old School support is based on faulty premises and will lead toward predictable and preventable problems, so too Sustainable Support is grounded in realistic premises. In this chapter we'll take a look at each functional premise, and then their practical results: the system works if you work it.

Premise 1: Technical Support is Primarily Communication.

William and Suzette are a newlywed couple. William loved early on in their marriage to spend time down at the public gym playing pickup basketball. A troubling pattern started to emerge, though. After intense (and sweaty!) hoops games, William would come home and quickly disrobe to hop in the shower before heading to work.

In his haste, he'd aim his sweaty gym socks for the dirty clothes basket and often miss, landing in a splat on Suzette's pillow.

At first the conversations were civil, and then (as newlywed fights tend to be) there was some passive-aggression followed by "making up" and mild escalation, as the problems would continue.

William would promise to do better, and then fall back into bad habits, and over time the issue got more and more serious, simmering in Suzette's mind.

This went on for two years, until one day in a fit of rage Suzette blew up at William, screaming and shouting.

William, taken aback by his normally-patient wife's outburst, said "we need to go to marriage counseling!"

"I don't need counseling, you idiot! I need you to just stop putting your nasty socks where I lay my head to sleep!" Suzette screamed.

Dismayed, William took it upon himself to go by himself to marriage counseling.

Imagine with me as William sat down with this professional counselor, answering the "what brings you here today" question with the following:

"My wife blew up at me for no reason, and called me an idiot (and worse!)"

This could be the world's best marriage counselor and it would not matter. Without a ton more context there's no hope for getting to the bottom of the issue, at least not quickly.

Tech support is like users taking their relationship with software to counseling. Usually the software can't talk, so the best "counselors" are the ones who are able to get the user talking as much as possible.

Until you are able to isolate the actual reason that the software is having issues, you're as stuck as the fictitious marriage counselor looking down the barrel of William and Suzette's communication problems.

Technical support is *communication* and the better your agents can be at distilling what's not being said from the context of what is being said, the happier they will be, not to mention the users.

In part one we mentioned the faulty premise that support technicians are viewed as "junior junior developers." Here's what they really are: a marriage counselor meets a private eye meets a linguist. Of course, the more knowledge overlap they have with development principles the easier it will be for them to get to the bottom of individual issues, but given the choice between someone who can *do* the debugging and someone who can *explain* why debugging should be done, the choice is clear. Since Technical Support is primarily communication, get the communicator.

That's the first premise undergirding Sustainable Support: it's communication. The more informed agents are, the better they are going to be at doing technical support over the long haul.

Premise 2: Technical Support fills in between departments.

Take a look at your organization chart at a typical software company, and at best you'll see Technical Support as a subset of the product team. Worse, if it's a financial P&L sheet you'll always find Technical Support in the "costs" category, which makes it easy to see it (and the team members themselves) as something to minimize and avoid.

Instead of leaving them in their own corner, Sustainable Support moves them more centrally into the product team. Technical support agents are the best Quality Assurance agents, the best documentation writers, and the best ones to ask those "silly" questions (like "is this something our customers will actually use?") during product strategy meetings.

It doesn't stop there. Technical Support agents can also provide irreplaceable insight about marketing campaigns, or financial insights, or the best ways to spend advertising budget.

The more you can cross-populate teams throughout the organization with support agents, the better. This is a person who regularly talks to customers, knows why they love (or hate) your product, and can provide more customer insight than even the highest-paid consultant. You just have to treat them like it. That leads directly to the last premise:

Premise 3: Technical Support is a Skill Position

When I was an undergraduate at the University of North Carolina, my time overlapped with Julius Peppers. He was a freakishly gifted athlete on both our college football and basketball teams. At 6' 7" tall and nearly 300 pounds, he was tough to miss.

His football position was Defensive End, meaning his job was to line up across from equally large humans who were trying to prevent him from getting to the quarterback or running back with the ball. His job was tackling, and he was very, very good at it.

I was always mildly amused when football commentators would refer to "skill positions" like Running Back, Defensive Back, or Wide Receiver. I know what they intended, but to call Julius Peppers anything other than "skilled" remains laughable to me. It's not the same skill as catching or carrying a ball, but his craft was undeniably "skillful."

Technical support agents get treated like "unskilled positions" at too many companies.

It's beyond time to stop treating (and paying) technical support agents like a drain on company resources. One of the only ways to compete left in business is customer service. Technical Support agents who treat customers with respect, kindness, and excellence are where the revenue lives.

The skill of communicating with customers in ways that leave them feeling heard, understood, and helped might not be the same as the skill to design pixel-perfect layouts or to code an interface from scratch, but it's still a skill, and deserves to be compensated as such.

Results of those functional principles:

Just like the faulty premises will always result in faulty outcomes in Old School Support, the functional premises properly applied will result in functional and sustainable outcomes:

1. Inbox zero is no longer a goal, it's a red flag.
2. Users are empowered.
3. The user's voice gets heard.

1. Inbox zero is no longer a goal, it's a red flag

In Old School support there's a risk that inbox zero becomes an unhealthy obsession, leading agents to cut corners and manipulate numbers to obtain it. In Sustainable support, agents have a much healthier relationship with the inbox: more tickets is a positive thing.

Since there's no such thing as a product so good that it doesn't need support, an empty inbox is a sign of one thing: customers are not using the product.

Here's a helpful mantra when opening the inbox: if people care enough to be mad at our product, that's a good thing. If they didn't care, they'd leave. So the fact that they are here and angry is better than not here.

Sustainable Support creates agents who don't shy away from a full inbox. An agent who has true agency in the company is acutely aware that fewer tickets is a problem.

To be clear: there are valid reasons and seasons to want to stop seeing certain types of tickets. We don't want more tickets just for the sake of more tickets any more than we want fewer tickets. There's the key to this result: the number of tickets is simply the wrong metric to track, and wrong information to fixate on.

A Sustainable Support agent or manager will be asking "what kind" instead of "how many" when they look at a support inbox, regardless of how full.

2. Users Are Empowered

In the Old School Support model, the user is alone on an island, alongside their problem. They are forced to be their own advocate against the problem.

In Sustainable support, with the teams arranged as "All Y'all" against the problem, the user themself is empowered to be a part of the solution. While Old School support reveres the technical support agent as the *One to Solve the Problem*, the new model rightly applied puts the user squarely in control of their destiny, empowered by the wise guidance of the support agent.

Old School support tries (and fails) to make the technician (or your company) the main character of the story. Sustainable support makes the agent the guide.

The Technical Support agent is...

...more Gandalf, less Frodo.

...more Fairy Godmother, less Cinderella.

...more Dumbledore, less Harry Potter.

...more Aunt May, less Spider-Man.

...more Miyagi, less Daniel-son.

...more Phil Jackson, less Michael Jordan.

Once you apply these Sustainable Support principles, The user or customer starts to see themselves as a part of the solution, rather than a passive non-agent waiting on rescue.

3. The user's voice gets heard

If technical support is communication, support agents are constantly being trained toward being the customer's voice on all the various teams, leading to a mindset of "every ticket is an opportunity to improve something."

In this way, Sustainable Support gives the user a real voice into product, documentation, and internal processes.

The more agency you give agents, the better this works.

There are three categories of change to look for:

1. Documentation updates

4. Product UI/UX updates

5. Quality Assurance updates

Every. Single. Ticket. ...can have an impact in one of those areas. The more senior a support agent, the more spots they can find to improve the customer experience.

Documentation Updates

Some quirks of the product need a special section in the documentation, to prevent the same old question over and over. A Sustainable Support approach to that would be to take the additional 15 minutes to craft one excellent answer and put it in the docs (coupled with a text-snippet response that introduces that new doc) and you save countless hours of time responding to that same question repeatedly.

Product UI/UX updates

Some quirks of the product need to be addressed in the product itself. The engineers who made the product often don't use it the same way customers and end users do. Without the expert help of a technical support agent, they likely have no idea how the product can be improved. Is the settings page unclear or confusing? Are people asking for refunds and leaving because of a specific lack of feature? These are all ways that a technical support agent (immersed in Sustainable Support principles) is uniquely positioned to be the voice of the customer. Some tickets should result in making the product itself better.

Quality Assurance updates

Finally, every time there's a bug or conflict with the product, a Sustainable Support mindset means that we not only escalate the replicated problem to the product team for a fix, we create a test or Quality Assurance check that means that same problem can't sneak back in without being noticed. Support technicians when properly empowered are the ideal agents to help shore up engineering and release processes. It's the only chance that development can be truly customer-driven.

Far from an antiquated "the customer is always right" mindset—they often are not right and need to be told that—the Sustainable Support model positions the Technical support agents as experts that are able to sort out and evaluate customer feedback into helpful categories. The longer a support agent sticks around, the more valuable their contribution becomes. That's why it's critical to build your support team sustainably.

The overarching effect of building a support team on Sustainable premises is that the customer gets a real seat at the table. The more you can put a Technical Support voice into every meeting, the better all of the teams (Marketing, Product, Quality Assurance, and even Finance!) get.

CHAPTER FIVE:

TECHNICAL SUPPORT AGENTS AS AMBASSADORS

In the remainder of this book we will look at practical implications and strategies to build that team. Before we do that, a quick overarching helpful analogy:

Your Support Agents are Ambassadors.

Ambassadorship is the role of representing one country or organization in another, typically through an appointed ambassador. An ambassador serves as a bridge, working to maintain good relationships, negotiate agreements, and communicate their country's or organization's interests while respecting the laws and culture of the host nation. Their primary goal is to foster understanding and cooperation between the two sides.

Sustainable Support systems only work because the Support agent, supervisors, and management all view the role of technical support through the lens of Ambassadorship.

Let's take a closer look to identify the similarities between effective and Sustainable Support and effective geopolitics via ambassadors:

Ambassadors are Empowered and Connected

The ambassador speaks with the full authority of the government of their home country. They are connected, and empowered. Without true connection and empowerment, the entire arrangement falls apart. An ambassador without true connection and empowerment is a propagandist externally and mere decoration internally. Sending nations ought to either actually connect the ambassador to "the room where it happens" or just not waste their time calling them an ambassador.

Ambassadors are Disarming and Resourceful

One of the key reasons that the entire concept of ambassadorship exists is to de-escalate conflict. Ambassadors add an element of relationship to the sometimes tense interaction between nations. Having a personal relationship with a connected and empowered person allows for that disarming.

The best ambassadors are the ones who can do the de-escalation without compromising on principles. Finding ways to make everyone happy is a skill that involves creative thinking and creative action. A personal relationship is a powerful thing, and sending an ambassador to make those relationships to foster goodwill pays off during moments of conflict. Ambassadors don't just need to be empowered, they need to be resourceful.

Ambassadors are Personal and Winsome

Ambassadors get to know the kids' names and sit through their dance recitals. Ambassadors drop off a meal when someone has an unexpected hospital stay. Ambassadors grab a drink after work to get to know people. They are the type of people you enjoy spending time with: positive, encouraging, and winsome.

Ambassadors are well-compensated

I've avoided talking about the finances to this point, but now before I show you the actual job description I should pause to acknowledge what I hope has taken place in your mind as you've read through these first two parts: a commitment to

Sustainable Support changes the compensation level of an agent.

The technical support "game" of having entire tiers of agents that only exist as a first line of unskilled, unresourced, and unconnected defense and are paid a fraction of what the next tier gets is fundamentally incompatible with Sustainable Support. The first person a customer talks to is an expert in a Sustainable Support system.

That said, part of the reason tech support is so cheap is that it's a soul-sucking job that has brutal churn and turnover. When you make the job significantly less likely to burn people out, and provide agents with training to handle difficult situations in proactive ways, you end up saving tons of money on that churn-and-burn cycle. It costs less to keep team members happy than it does to train new folks to become experts.

Ambassadors are NOT Adversaries

Ambassadors are allies, friends, co-laborers, and teammates. It doesn't matter who they are talking to: they are representing the needs and desires *of the other party*. In both geopolitics and technical support this looks like a continual reevaluation of priorities and resources around making sure that everybody is being understood, and represented well.

A support agent given true agency as an ambassador is a key piece of the strategy to build Sustainable Support that benefits the company, the customer, and the agent.

SUSTAINABLE SUPPORT

Part Three
How to
Create a
Sustainable
Support
Environment

INTERLUDE: STRETCH BREAK

Ever been on a road trip, and just before you reach the big city you decide to take a quick break to stretch your legs? That's where we are in this book.

There's not many wasted words or repetition from here: things are going to get intensely practical, culminating in me literally giving out the system I use to train new team members.

Before we do that, a quick refresher and break like a sip of cool water and a chance to walk around and stretch your legs.

To this point in the book we've looked at what Old School Support is (and why it's ultimately no good!) and at what Sustainable Support is. All of the practical steps and tactics are rooted in the premise that Sustainable Support benefits the company, customer, and support agent.

Any lapse in Sustainable Support tactics can be traced to a breakdown in the underlying principles we've covered already:

→ Agents are not engaged.

→ Customers are asking for refunds, or leaving negative reviews.
→ The product team is isolated and standoffish.

Every bad outcome can be traced to a principle.

The goal is for everyone involved from leadership to the front line agents to be able to spot those breakdowns using a mental model to run their replies through. Like anything, the key to mastery is practice and feedback. So how do you as a leader set up a system that's transferable to every member of your team?

Part three of the sustainable Support system will give you practical, repeatable steps to take, practice, and make your own. This system has worked for me, but that doesn't mean it's going to work for you. Instead, treat it as a starting point to think through how you can apply it to your specific context.

All gassed up, and legs stretched? Let's power through the end of the book.

CHAPTER SIX:

LEADERSHIP LESSONS

When my kids were first learning to ride a bicycle, I had all sorts of incorrect expectations. I thought that I'd show them how to do it, then strap their helmet on and return moments later to have a video montage of them victoriously stepping onto the pedal, swinging their leg victoriously over the saddle, and sweeping down the gentle hill in the neighborhood park.

It turns out, as I quickly remembered, nobody is ever good at anything the first time they try it. They had to fail a few times on the soft grass, then get it right, then fail again, then try it in a slightly different context (flat ground! up hills!) And fail there a few more times. Improvement in cycling is almost never truly linear.

"Nobody is ever good at anything the first time they try it" eventually became a bit of a mantra that I've used with those same children (and it applies to more than bicycles!). The only way anybody has ever gotten to expert level in any field is by willingly stepping into being terrible at that same activity.

The best plumbers have flooded a basement.

The best electricians know what it feels like when you forget to turn off power to an outlet before you take a screwdriver to it.

The best carpenters have hit their fingers with a hammer more times than you and I (probably 10x more times!).

A good leader (just like a good parent) doesn't shield their team from failure: they make it a safe place to fail forward. Your job as a leader in Sustainable Support is to provide the grass to fall on, and then show the team it's not so bad.

To that end, three steps:

1. Make it personal

6. Make it objective

7. Make it continual

Step 1: Make it Personal

Have you ever used one of those compressed "log" fire starters to get a campfire (or fireplace fire) going? It's a genius bit of time savings. Gone are the 15 or 20 minutes you have to spend making sure that the fledgling flames have enough fuel and oxygen to ignite small kindling first, and then gradually larger sticks until the largest logs are ready.

With the fire starter log it's one flick of a lighter, and 2 minutes later there's a fire hot enough to skip straight to the largest sticks in your pile.

I've never met someone who gets the principles of Sustainable Support right the first time. No manager does it, no agent does it, and I didn't do it right the first time, or even the first 10 times.

The "fire starter log" to accelerate your team toward Sustainable Support? Using yourself as the negative example. You're *going* to mess it up, so you might as well cash in on showing your team that this is a safe place to mess things up.

Make it personal: you do it, and demonstrate it.

The best way I've seen to do this is to conduct at least one Customer Response Review (or CRR) per week.

Customer Response Reviews

As the name indicates, a CRR is a chance to take a ticket (or chat, or call) and dissect it, mining it for clues on how you did. Practically, a CRR is a public walkthrough of a ticket where you did a particularly bad job (or a particularly good job, but that can backfire) lining out what you understood from the customer, how well you did articulating the solution, and what you could have done better to get to resolution faster.

A good CRR is essentially verbally processing of things like

- "Here's what I was thinking at this point"

- "Here's why I asked this followup question in the way I did"

- "Here's what I think I would do differently if I could go back and do it again."

We'll cover some objective metrics you can use during CRRs in chapter seven, but the point of step one is that leaders make it personal and use "I could have done this better..." kind of language.

It's certainly OK to use "good outcome" tickets, but I've had much success reviewing the tough ones where I see immediately how I could have done better.

The value of the psychological safety of being able to say "we're all getting better every day here" is hard to overstate. The CRR is one of the keys to a Sustainable Support team.

The more you do CRRs, the easier they'll get, but also the easier it'll get to review others on the team. The end goal is to create a collaborative environment where it's OK (and encouraged) to not know things, so that the team can grow together.

Step 2: Make it Objective

Feedback that is vague or based on emotion is better left unsaid.

Zig Ziglar famously said "If you aim at nothing, you'll hit it every time." If your feedback is not based on some objective and defined standards, you can guarantee that you'll be giving that feedback again the next time you see the problem you are trying to address.

We'll include specific swipe-able metrics to see what this might look like, but here's a principle that should guide you as you set the metrics you're going to "grade" on: If feedback is not objective, there's no way you can teach someone else to give that feedback.

For longer than I would care to admit, I was frustrated with supervisors not seeing the clear evidence of how a reply could be better. Upon further inspection I ran against the harsh reality: they couldn't give good feedback because I hadn't clearly defined both the metrics and what constitutes a certain score for each metric in a way that wasn't overly dependent on some subjective judgment or personal opinion.

Agents hearing "that reply could be better" won't make the next reply better.

Instead give specific and objective things to change, coupled with the principle underlying the change. Here's what that might look like:

"You said 'I'm sorry' here when you should have led with confidence. Leading with 'I can help with that' will help both you and the customer enjoy this interaction more."

The more the feedback is tied directly to the objective metric, the more anyone will be able to see how to train a new technician on giving feedback. So it's triply unhelpful (to the customer, to the agent, and to the supervisors) if the feedback and the metrics are not objective.

Step 3: Make it Continual

A few months ago I was describing to a friend (who manages teams in a non-tech industry) how often we do response reviews with my team and he was flabbergasted. "If I reviewed the performance of my team members that often they'd leave immediately!"

That was my first indication that what I am about to say may ruffle some metaphorical feathers. So I'll start bluntly and walk back toward agreement:

Every technical support agent should be receiving 2-3 points of constructive feedback per week, as a minimum. I'll get into what exactly that looks like soon, but the bare starting point for building a Sustainable Support team is 2 times per week being reviewed.

Let me be clear: these are not some sort of yawning middle management exercises where we check a few boxes to make sure our backsides are covered. Obviously they can be used as an HR lever to accelerate getting rid of perpetually low performers, but that's **not at all** the primary reason you should do them, or how you should phrase it to your team.

The goal of the review is calling the agent (and therefore the whole team) to excellence and job satisfaction (with a side of thrilled customers that generate 5-star reviews and make the leadership team very happy).

Let me say that again and slightly differently, in case it was unclear: the point of interaction reviews is agent satisfaction almost as much as it is customer satisfaction.

We're far too stuck as an industry in an Old School mindset that puts unnecessary existential weight on "performance reviews" so it's important you communicate that a CRR is not primarily a HR tool, it's primarily a tool to reinforce difficult-to-internalize principles in a way that everyone involved from customers to agents to supervisors can know with certainty that they are hitting the mark of Sustainable Support.

Said again, and even more pointedly: Personal, objective, regular, and honest feedback from happy supervisors makes happy agents who make happy customers.

Sustainable support is more fun.

One more Parenting analogy

My boys are the only male grandchildren on my wife's side of the family. When our oldest two were 3 and 1, we would visit family out of state, gathered in homes not designed with young child safety in mind. All the girl-parents would marvel at how my boys would seemingly instantly find the most dangerous way to scale the outside of a staircase, or test to see how much it hurt to fall off the trampoline.

Aghast, our well-meaning relatives would insist that we do something to prevent the boys from climbing.

"I am not going to stop them climbing. Instead, we've taught them how to fall correctly. Don't brace your fall with arms, tuck your head in and roll when you hit the ground, etc."

You can't stop my toddlers from doing anything that might hurt. Neither can you learn Sustainable Support without failing and doing it wrong from time to time.

You can create an environment where failure is not punished, and while not encouraged, it is not treated as a completely unexpected event.

Lead by example with your team. Make it personal, objective, and regular, and teach them how to fall down in a way that they can then get back up.

Part Four

Steal My Playbook

FINAL INTERLUDE: JUST THE SWIPE-ABLE STUFF

So how do you actually do a Customer Response Review? What do you actually say when you look at how your team is doing and want them to move toward Sustainable Support?

In Part Four we'll look into how to make CRRs indispensable and not a waste of time (scoring, philosophy, and specific metrics), we'll take a look at the value and process of postmortems for bad support outcomes, and we'll look at how you and your team should handle escalation of both technical issues and angry customers.

The goal is to create a system where agents know how to unstick themselves, supervisors become professional unsticking balls of encouragement, and directors and management are freed up to keep the system up and running.

CHAPTER SEVEN:
CRRS AND THE SCORECARD

Before you can expect an agent to grow, you need to show them what a perfect score is, and then point (repeatedly) at that standard. There are 4 areas that every ticket, call, chat, or customer interaction can be graded on.

Before we look at the metrics it's helpful to know how scoring works.

Scoring CRRs

Each individual metric is scored on a 3 point scale.

1 indicates a problem. An agent should feel nervous on some level with every score of 1 on an individual metric. The only agents that shouldn't feel nervous with 1s are brand new agents, the first time they get a 1 for that metric.

2 indicates no problem. An agent should feel no nervousness about a 2.

3 indicates an exemplary answer. An agent should be proud of a 3 and want to share it.

Put differently, we want for the scores to communicate this:

Keep getting a 1 on this metric and you'll likely be moving toward disciplinary action. It's not good enough.

Keep getting a 2 on this metric and all is well. There's no need to worry about disciplinary action. It's good enough.

Keep getting a 3 on this metric, and you're positioning yourself for a raise or promotion into leadership. It's exemplary, the type of thing we should want every agent to see.

CRR Metrics

Before I get into the specific metrics, a quick filter that applies to all of them: show don't tell.

For each metric, supervisors need to evaluate how well the user would rate the agent having *demonstrated* it. It's not enough to say it, and in fact sometimes saying it has unintended effects that are worse than not saying it.

Chief among triggering phrases in technical support: "I understand." There's almost no case where telling a user that you understand is a net gain.

When an agent says "I completely understand…" there's a really good chance the only thing the user is thinking is "no, you don't." *Show them* you understand and let them be the one to tell you.

It's like calling something "easy." When you *say* it, the user rolls their eyes. When you *demonstrate* it, the user will use the word "easy" in the glowing review they post online. Let customers call it easy and simple. You just describe how to use it.

OK, on to the metrics. Grade every CRR on every metric: Expertise, Ownership, Articulation, and Proactiveness.

Expertise

Expertise is answering the question "Do they demonstrate technical grasp of the problem and surrounding issues to the user?"

It's not only important that the agent knows what the problem is (or how to isolate and find the problem), they have to communicate that knowledge to the user helpfully.

A 3 on expertise positions the agent as an expert on the product they are supporting, and all of the ways that product interacts with other products in the space.

By the end of a 3 response, the user is confident that the agent is not only highly knowledgeable about the product, but willing to share that knowledge!

Expertise quick tip: confidence is key. Even if they're not sure what exactly the problem is, an agent demonstrates expertise by projecting confidence that the moment they are able to isolate the issue, they are going to be able to get it fixed.

Ownership

Ownership is answering the question "Do they avoid passing off responsibility to others to solve the problem?"

A 3 on ownership positions the agent as the ambassador navigating the conflict between products, or support desks, or the user and the product.

By the end of a 3 response, the user feels like the agent will not stop until the problem is resolved.

Ownership quick tip: the goal of the interaction is resolution, but don't mistake that to mean "the bug has been reported" or anything that simple. An agent that truly owns the interaction has communicated timelines, given workarounds, and taken care to make sure that the user leaves with an immediate plan.

Articulation

Articulation is answering the question "Do they communicate clearly, with helpful tone, and attention to detail?"

A 3 on articulation nails the tone (more on tone in the Tone Guide), pays close attention to details in the customer's word choices (and often will ask clarifying questions based on those choices).

By the end of a 3 response, the user feels like they have a wise friend at the company.

Articulation quick tip: agents are not robots, and every interaction is a chance to wow someone with small touches like humor, personalization, and clarity.

Proactiveness

Proactiveness is answering the question "Do they utilize all opportunities to maximize long-term benefits of this interaction?"

A 3 on proactiveness finds something in the interaction that can make the company better, the product better, or the documentation better, and not only solves the problem for the user, it seeks to prevent others having that same problem!

By the end of a 3 response, the user feels like the agent is their secret weapon to get the direct ear of the product team, and the user understands more deeply how product decisions are made.

Proactiveness quick tip: every single interaction can have an impact on the company. Taking the extra 2 minutes to thank the customer for help making the product or documentation better can be the difference between a refund and a 5-star review.

Chapter Eight:
Tone Checklist: is it C.R.E.W?

(This chapter's content is taken directly from what was originally the GiveWP support manual, later incorporated into the StellarWP support manual. I wrote this alongside Matt Cromwell, and this content was released under a GPL copyright license.)

We are our customer's wise friend, not a corporate policy enforcer or policy defender. We want our customers to succeed, and we are the guide to help them when they need it. Our customer is James Bond, we are Q: equipping them with the tools, resources, and knowledge to win the day*. To that end, our tone is marked by these qualities (with the helpful acronym CREW):

- **C**onfident, not Apologetic

- **R**esults Oriented, not Argumentative

- **E**ducational, not Haughty or Overly Technical

- **W**elcoming, not Thankful

The best support technicians know their biases and tendencies. For example, if a technician is prone to argument, they should take special care to make things results-oriented. It's helpful to do a postmortem of all negative rating interactions to see which part of the tone guide broke down in the process. Almost all negative ratings (even the unfair ones where they rate the technician negatively because of a lack of feature) can be traced to a breakdown of one of the four points in CREW.

Pro Tip: *Slow Down.* There's no substitute for proofreading when it comes to tone. It's a best practice to take the extra 30 seconds to re-read drafts checking for each of these items before sending the message, asking the question "is this CREW?" and modifying the message before sending it.

Confident, not Apologetic.

We are proud of our product and company and don't apologize for things that are not our fault. We empathize with our customers and don't waste time defending our products or ourselves, but we never apologize for things beyond the scope of our (team's) control.

Things to apologize for:

- "So sorry for my slow response to this issue. The fastest way to resolve the problem is _"

Things you shouldn't apologize for:

- "I'm sorry for the theme conflict you are experiencing."
  Instead: "We work very hard to ensure that our products are compatible with the vast majority of the themes out there. It looks like our product is not playing nicely with your particular theme. Here are the steps to resolve this:"

- "I'm sorry for this bug"
  Instead: "This issue has been urgently escalated and our developers are working as I type this to release a patch. I'll let you know as soon as I hear anything from them."

Results Oriented, not Argumentative

Many of the folks who reach out to support are frustrated, having attempted to solve a technical problem for potentially hours before reaching out. In their ticket or response they can resort to attacking us, lashing out at the product or team, or getting distracted by inconsequential details that don't move the issue toward resolution.

We do not respond to their frustration with a defense of our actions, our team, or the current reality.

Instead, we seek to find the source of their frustration, and resolve it.

For example, let's say a customer replies to you saying "Your team is incompetent".

- **Ineffective response:** "Actually we're quite competent. I have a degree in {Something}, have spent the last {however many} years in customer support, and thousands of people are thrilled to use our products daily"

- **Effective Response** "I'm happy to get to the bottom of this with you. Other users who have had this problem resolved it following these steps:"

Functionally, we simply ignore the insult, and then prove them wrong by resolving the problem.

We never stand for sustained mistreatment of our team by customers, but can understand and empathize with frustration. While we won't work with customers who are repeatedly abusive towards our support team, it is important to us to step into their shoes and be on their side to resolve the problem.

Most of the time resolving the problem results in a change in tone from the customer. When it does not, those situations get escalated to supervisors.

Educational, not Haughty or Overly Technical

We are experts from a technical standpoint. We understand hooks, functions, PHP, and JavaScript, and are able to resolve problems. There is a temptation when talking to customers to show them how much we know by using jargon, insider-speak, and overly technical language.

Far from making the customer feel better, overly technical jargon makes them feel incompetent and ignorant. If our job is to be the Q to their James Bond, we have to always set them up to understand the things we are showing them.

For example; let's say a user had a Javascript error that was making things difficult for our plugin.

- **Ineffective Response:** "The problem is a script loading in the header that's throwing a jQuery error preventing the rest of the form from loading"

- **Effective Response:** "A file that is loaded by {Example} plugin is not playing nicely with one of our files, and that conflict is preventing the form from loading correctly. Here's how I recommend confirming and resolving the issue..."

It's helpful to imagine scenarios where you are in similar situations. One example is the doctor's office. Imagine going to a doctor for a problem with your daughter's development.

- **Ineffective Response:** "The issue here is acute craniosynostosis of the posterior temporal suture"

This response, littered with medical jargon, serves no purpose other than to demonstrate to the listener that the speaker has a large vocabulary, and possibly a firm grasp on a medical condition. In a strange twist, the more technical language actually proves less about the competence of the doctor than simple and understandable language would.

- **Effective Response:** "You know how the skull has gaps (soft spots) in it when you are born, that you can feel? The issue for your child is that one of those gaps, specifically one of the ones on the back of her head, closed prematurely. Here's how we go about fixing that..."

If you can't explain it to a ten-year-old, go back and rewrite it.

NOTE: Avoid being patronizing to developers and agency folks who reach out. If they reference the developer console and provide a specific line of code where they are having a problem, treat them with respect, and educate them specifically about our products.

Welcoming, not Thankful

Thankfulness, especially as an opening, is functionally white noise to the customer. Like an automated greeting that says "Your call is important to us..." the line "Thanks for contacting support" is ignored at best, or ridiculed at worst.

Instead, we are excited and welcoming. Taking the few extra seconds to find some common ground with a customer and personalize the welcome is a difference between a customer being on the defensive (because they are prepared to meet the policy enforcer) and being delighted that they have found their wise friend to help them win the day.

For example, how you set the tone with your first sentence in your first reply makes a big impact.

- **Ineffective Opener:** "Thanks for contacting our support"

- **Effective Opener:** "It's so great to help raise awareness for Autism and special needs children. I've seen this technical issue you're experiencing before, and look forward to getting you back to doing what matters: helping the kids."

NOTE: Don't fake it. If you are not excited about their organization, skip straight to "You've come to the right place: we'll get this technical issue sorted for you in no time." Lead with being welcoming, even if you can't lead with excitement for their site.

When people go to tech support, they brace themselves for being made to feel stupid. Our job is to welcome them, and explain that we've been here before.

Chapter 9:

Closing the Loop: How we Expect, Prevent, and recover from Failure

We covered earlier that thus far in my extensive experience nobody (even myself!) has ever gotten Sustainable Support right on the first try. It requires practice and a careful posture of continuous improvement. That said, there's still ways that we need to see better outcomes starting soon after we start consciously moving toward Sustainable Support.

Nobody can afford to set up a new system that fails to deliver for 6 months.

To that end, the final things we need to cover are proactive and reactive methods for making sure you capitalize on those opportunities for learning.

Proactively, the method for preventing failure is to promote and foster a healthy culture of communication internally so that agents and supervisors know how to get unstuck quickly.

Reactively, once failure happens we have to objectively and carefully respond.

Work In Safely Public Spaces

There's no better shortcut to expertise than safely public (and often borderline awkward) ignorance. Just as with CRRs, this starts from leadership modeling "things they need help with" as much as possible.

"Safely Public" is different from being *truly* public. Few aside from some borderline sociopathic folks are willing to be demonstrably ignorant in a truly public space (where customers and even trolls could capitalize on perceived weakness.)

As a manager, if there's not a space where support agents can feel safe to be transparently ignorant, Sustainable Support will remain elusive.

Perhaps predictably by this point in learning the system, it starts with leadership. The faster you as a leader can model safely publicly ignorance, the better.

Confusing ignorance with its more sinister counterpart incompetence often leads to a critical misunderstanding that I should get ahead of.

There is a difference between incompetence and ignorance.

I am *ignorant* of the names of ballet dance moves (or even if they are called "moves," if I'm being transparent here). At the same time, I'm an *incompetent* ballet dancer.

Incompetence is a judgment on my abilities—flexibility, stamina, body control, capacity for growth physically. Ignorance, by contrast, is only a judgment on my exposure to and understanding of the concepts.

Ignorance can be fixed, thankfully easier than teaching me to pirouette—the only "move" I don't have to use spellcheck for.

Safely public ignorance works like an immediate and irresistible invitation for correction: people with knowledge can't seem to help themselves when it comes to filling up other people's knowledge gaps.

The fastest way to fix my ballet ignorance would be to waltz into a room filled with ballet dancers or former dancers and just stumble around saying things like "I'm not sure what this is called" or "how many foot/arm positions are there?"

In technical support, things are constantly changing: new versions of software, new processes being refined, and new challenges from competitors and the market.

There's no shame in being ignorant (within a team channel or room) for the express purpose of becoming effective at technical support.

Private ignorance grows until it becomes hidden incompetence, which acts like a time bomb on your technical support team.

Safely public ignorance gets rooted out until it transforms into public expertise.

If work is exclusively happening in Direct Messages or 1:1 calls and emails, there's just about no way to catch and correct ignorance.

Quick aside, though: the only way your team can be publicly ignorant is if they are psychologically safe at work. So before you start berating them to be more ignorant in public, do it yourself, and be very vocal about how the goal is to move toward excellence, and that nobody is getting fired for not knowing things.

Like being introduced to someone at a party only to immediately forget their name, the earlier ignorance is made plain to other teammates, the less awkward it is to correct.

Here's the two things you can implement today in a safely public space: Escalation Notes and Postmortems.

Escalation Notes

Sustainable Support encourages a constant and safely public space for agents, supervisors, and management to constantly be adding product and market knowledge, answering support requests, and filling the role of customer ambassador.

One profound enemy of the entire system is agents getting "stuck" in technical problems or procedural questions.

Sustainable Support needs agents who know how to get unstuck, and managers and supervisors who are essentially unblocking machines, consistently finding ways to simultaneously train agents and set up systems and procedures.

When an agent gets stuck, the supervisor needs to be able to unstick them quickly, and to do that, they need to fully understand the situation as quickly as possible. Enter the multitool every team member should use: the escalation note.

Before I explain what escalation notes are, let's clear up what they are not. Without fail in training agents to use escalation notes, I've heard the same objection (and even had it myself): escalation notes feel like administrative busywork.

Escalation notes are not busywork. The more agents get in the habit of writing escalation notes, the more value those agents will get from them. I've gotten to the point now where they are an indispensable part of my workflow. Busywork wastes my time. Escalation notes end up saving me time.

So what is an escalation note? Briefly, it's a short but detailed written message designed to take someone (usually a supervisor) from uninformed to fully up-to-speed on a given issue.

Like an onramp to the situation, an escalation note provides a way for agents to bring others into the problem-solving process as quickly as possible.

Let's be very practical here. A good escalation note does these three things:

1. Immediately isolates and calls out what resolution would look like for the user, specifically.
2. Thoroughly walks through in detail the steps taken so far to attempt to isolate or replicate the issue.
3. Carefully points out what knowledge gap or process is blocking the agent from getting to resolution.

We have a template saved as a canned reply that guides agents through writing the escalation note.

Two pro tips:

- Agents can't add too much detail. The secret sauce of an escalation note is the breaking down of steps into nearly-comic simplicity.

- Agents will (almost) never start the note early enough in the process of getting stuck.

Break it Down

If I were a new car owner just learning how to drive, and I awoke one morning to a car that won't start, here's what a bad escalation note looks like:

1. I tried to start the car, and it didn't work.

Here's a better note for the same problem:

1. I made sure that I had the right key.
2. I opened the driver's side door and sat down.
3. I checked to make sure that the fuel gauge is properly set to E for "energy" and not F for "fail."
4. I pressed the start button.
5. The car made a sound like it was trying to crank, and even fired once, but then immediately turned back off.

You probably spotted the problem. As a new car owner, I misunderstood right there in point 3 that

"E" is actually for "Empty" on the fuel gauge. My fix is simple, and involves the red gas can in the back corner of my garage.

Without detail, my ignorance of the markings on the fuel gauge would have likely gone unnoticed, and the supervisor reading my escalation note would have to spend significant time getting up to speed. Simply verbalizing this detail made the note effective.

Start Sooner

Practically speaking, agents should start writing an escalation note sooner than they think.

Why? The longer agents metaphorically bang their head against a technical problem, the more frustrated they get. By the time they're truly stuck, if my experience is normal, they're essentially a cave person who's lost their ability to use words: "ME PRODUCT BROKEN. YOU FIX."

At that point, the very moment they realize their need for an escalation note, they've lost the ability to write it. All the agent has left are expletives.

So start earlier.

Escalation notes end in at least one of three predictable outcomes:

1. The problem resolves itself

2. The problem becomes immediately solvable by others

3. The problem becomes a catalyst for solving systemic or product problems

Self-resolving Problems

In high school and college I made some spare money by teaching guitar lessons to kids from my church. Once as I was in the midst of explaining where a student should place their fingers on the fretboard to make a popular chord, the student ambushed me with a two-part question: "how do I strum? What strum pattern should I use?"

I've played guitar since I was very young. So young that I don't remember what it was like to learn strumming. Sitting there with my eager students I had to grapple with how to explain a thing that I do subconsciously.

Escalation notes use the "explaining" part of the brain. Suddenly instead of "doing" the thing, agents are "explaining" the thing. Explaining uses a different (and usually more level-headed) part of your brain, and pulls you out of the process enough to be able to see more clearly.

That leads to what used to surprise me about escalation notes: Just by thinking about escalating something, all of a sudden I don't need to! I love problems that resolve themselves.

I end up not even sending easily 50% of the escalation notes I write. And no, in case you're wondering, I've still never thought "that was a waste of time!" or "I started that note too early."

Easily Solvable Problems

If I were to pay a local golf professional to go with me to a driving range for some lessons they'd likely be able to, within a matter of minutes, isolate why every time I use a driver the ball aggressively slices off the club, predictably out of the fairway to the right.

Why can golf pros isolate the problems in my swing? Because they've shanked more golf balls into more woods, lakes, and hazards than I have, and they've watched 1000 more people swing a club than I have.

A detailed escalation note sets up a more experienced agent to immediately be able to isolate and potentially even resolve a problem.

Important, and at the risk of redundancy: without details, the "pro" is not going to be able to help your "swing." Escalation notes must have details (even ones you don't feel are necessary!)

An Open Door to Solving bigger problems

Lastly, if the escalation note doesn't lead to a self-resolving problem or a colleague-resolved problem, it may lead to you having a way to fix product problems in the documentation, or the product itself. There's nothing a developer likes better than clear steps to replicate an issue.

When an agent has a Sustainable Support mindset, they are going to be looking for ways to make the customer's voice heard on the product team. What better way than with clear steps to replicate an issue, or clear steps for a developer to be able to identify confusion? Escalation notes are truly a multitool.

With those three outcomes firmly in mind, allow me to reiterate the earlier point: Escalation notes should be done where everyone on the team can see them.

Public notes give more experienced agents the opportunity to help even if a supervisor is not available. Like a golf professional watching a beginner's swing, the more experienced agent is going to be able to immediately spot improvements in process, tone, and technical detail.

Everyone wins as long as folks are willing to be (safely) publicly ignorant.

Postmortems

The final piece of recovering from failure is the concept of postmortems: intentionally looking at those moments we don't live up to our ideals.

Any negative outcome should get a postmortem. A refund, a 1-star review, a bad customer satisfaction survey, or any other measurable bad outcome should be analyzed for insight into how it happened.

Just like escalation notes, there's real (and predictable) temptation to treat postmortems like administrative busywork. Also like escalation notes, I'm consistently surprised by how much value comes from looking back at an interaction with a few specific questions in mind.

Important: we don't do postmortems to assign blame, we do them to address systems and processes. If during the course of putting together the postmortem it becomes clear that the problem is in large part due to a specific individual, that should be dealt with separately and privately between the individual and their manager.

There's immense value in postmortems being publicly discussed, and even in cases where there was a private conversation, a team-wide awareness of the issues that caused the problem often reaps benefits.

Before I get into the specific questions I have supervisors process through on a postmortem, three pro tips:

1. Resist the impulse to mistrust the user.
2. Seek objectivity and recognize your own bias.

3. Find ways to tie outcomes to breakdowns of specific metrics.

Mistrusting the User

Because the user who gave the bad review is often factually incorrect and emotional, it's easy (if not automatic) to discount the entire substance of their complaint.

Many times the customer is unjustified in their anger and frustration, but that anger and frustration has naturally flowed from a combination of misunderstanding and lack of communication.

Our job as support technicians is to ignore the anger and deal with the miscommunication: we must get back *before* the anger and attempt to head it off at the pass. Sustainable Support is communication, and we have failed if the interaction has ended in a vocal 1-star review.

There are only a few cases where I've gotten to the end of the postmortem and truly come to the conclusion that there was nothing we could have done differently and the customer was just wrong.

Importantly: even those postmortems ended with action points to try and make the pre- and post-sales communications clarify in a way that would have helped with that misconception.

Seeking objectivity

It's natural to react to negativity by being defensive. Remember: being defensive is the only sure-fire way to make the interaction worse.

So before you even sit down to write out the answers to the postmortem questions, acknowledge that the more you can compartmentalize and dissect how your own bias toward defensiveness is coloring your response, the better.

The goal of the postmortem is to prevent the bad outcomes in the future, so it does no good to find ways that you were *right* in the past. Look for things that you or your team can change. Gather evidence for how you could do better, not for how the user could have.

Finding specific breakdowns

I spent more than half of the content here on philosophy for one mega-reason: I am convinced that every single bad outcome in Technical and Customer Support can be traced back to a breakdown of one of the philosophical points.

As you write the postmortem, you're looking through the lens of CREW, and intentionally aiming to root out those Old School Support impulses.

There's *always* some specific breakdown that you can link to the negative review.

Even those 1-star reviews where the user is 100% frustrated with the product and not the support (and will even say as much in the review) are a fundamental breakdown of a key principle of Sustainable Support: it's "All y'all" against the problem. The user shouldn't be separating you out into "product" and "support" teams. It was our job to keep that from happening.

The value of a postmortem is to surface the ways in which our team has not fully bought into the principles!

Swipe our questions

Here's the specific questions (we copy and paste this exact table into one doc for every postmortem)

Date:	Date of the bad rating/review
Associated Links:	Link to the bad review, the helpdesk ticket, any other pertinent documentation, etc.
Username or ID:	How you identify the user in your internal systems
Post-Mortem Author	Who is writing this postmortem
Original review	Copy the full review text here. (The goal is for users to modify the bad review, so it's nice to have a record of the original review)
Core Complaint:	In your own words, state the core of the complaint. Make this statement as objective as possible, and devoid of emotion.
History with this user:	Scour the helpdesk software and other systems for details: • When they purchased (if applicable) • How many other interactions they had before the bad review and the outcome of those interactions. • Detailed CRR of the most recent interaction ○ Use timestamps and names. Be specific. "At 10PM central on 4/14 the user reached out via email ticket. The agent responded 13 hours later (3 business hours) at 11AM."
Is this mostly about interactions with our team; or issues with the product?	This is a helpful question to ask, but it's important to note that the objective of asking it is not to give the support agent a pass because the customer is mad at the product. That's an Old School impulse to separate the support team and the product team.The goal is to find ways for the entire team (product, marketing, sales, support, finance) to work together to prevent these bad outcomes.
Action Items we should take to prevent more reviews like this:	• This should be a bulleted list of time-bound action items • Each bullet should be as actionable and practical as possible • Each point should be assigned to a specific team member • Bonus points for tying it to some specific metric or breakdown of that metric.
How we responded to the negative review:	Copy and paste in the reply to the review, and note any other actions or steps taken to attempt to repair the situation.

Postmortems should be curated by one individual, and then discussed in team meetings as safely publicly as possible.

Epilogue:

Where do you go from here?

That's it. I've emptied my bag of tricks onto the page. It's all here for you to steal, remix, and put to work.

I've personally watched 5 different teams implement this system, and I know it works.

Over the next decade, your company's ability to stand out, make more money, and stay in business boil down to how well your technical support team can truly take on the role of Ambassador.

It's well past time to put away the notion of technical and customer support as a cost, a line item to eliminate, and a role to automate. If you're in the C-suite, it's time to take a long look at the faulty premises holding back your teams.

For the rest of us, the only way to change the entire Technical Support industry is by changing one team at a time. Technical Support can be a fulfilling and highly valuable experience for customers, agents, and managers alike.

Here's my challenge: prove me wrong. Implement this system and then document how things become worse, customers become less happy, and email me about it at

ben@buildsustainablesupport.com

www.ingramcontent.com/pod-product-compliance
Lightning Source LLC
Chambersburg PA
CBHW071602200326

41519CB00021BB/6842